Animals on the Farm

Turkeys

by Wendy Strobel Dieker

Bullfrog Books

Ideas for Parents and Teachers

Bullfrog Books give children practice reading informational text at the earliest levels. Repetition, familiar words, and photos support early readers.

Before Reading

• Discuss the cover photo with the class. What does it tell them?

• Look at the picture glossary together. Read and discuss the words.

Read the Book

• "Walk" through the book and look at the photos. Let the child ask questions.

• Read the book to the child, or have him or her read independently.

After Reading

• Prompt the child to think more. Ask: Do you like to eat turkey? If you had a farm, what animals would you raise?

Bullfrog Books are published by Jump!
5357 Penn Avenue South
Minneapolis, MN 5419
www.jumplibrary.com

Library of Congress Cataloging-in-Publication Data
Dieker, Wendy Strobel.
Turkeys / by Wendy Strobel Dieker.
   p. cm. — (Bullfrog books: animals on the farm)
Includes bibliographical references and index.
Summary: "Turkeys narrate this photo-illustrated book describing the body parts and behavior of turkeys on a farm. Includes photo glossary" —Provided by publisher.
ISBN 978-1-62031-007-6 (hardcover)
ISBN 978-1-62031-634-4 (paperback)
1. Turkeys—Juvenile literature. 2. Turkeys —Behavior—Juvenile literature. I. Title.
SF507.D54 2013
636.5'92--dc23
                                    2012008426

Series Editor: Rebecca Glaser
Series Designer: Ellen Huber
Production: Chelsey Luther

Photo Credits: Alamy, 6-7; Dreamstime, 3bl, 5, 16, 20, 22, 23bl, 23ml, 23mr, 24; Getty, 4, 6, 8–9, 9, 14–15; Shutterstock, 3t, 3br, 10, 11, 12, 17, 20–21, 23tr; SuperStock, 1, 12–13, 14, 18–19, 23tl, 23br

Printed in the United States of America at Corporate Graphics in North Mankato, Minnesota

Table of Contents

Turkeys on the Farm

I am a turkey.
I live on a farm.

5

beak

Do you see my beak?
I eat feed, seeds, and bugs.

6

Do you see
my wings?
I do not fly.
I am too big.

Do you see the fuzzy poults?

hen

poult

These babies hatched last week.

Do you see the
brood house?

Poults stay
there.

It is safe
and warm.

tom

Do you hear him gobble?
He is a tom.
Hens do not gobble.

Do you see his wattle?

wattle

16

It is red when he is upset.
It is gray when he is calm.

Do you see
his feathers?

18

He puffs
them up to
look bigger.

Look at him strut.

Hens like a big tom that struts!

strut

Parts of a Turkey

feathers
Light, fluffy parts that cover a bird's body.

wattle
The skin that hangs down under a turkey's beak.

beak
The hard part of a bird's mouth.

claw
A hard, curved, sharp nail on a bird's foot.

wing
A limb of a bird that is covered in feathers.

Picture Glossary

brood house
A barn with heat lamps where poults stay warm and safe.

poult
A baby turkey.

feed
A mix of grains, vitamins, and minerals that is fed to turkeys.

strut
To walk tall and proud; toms puff up their feathers and walk around proudly.

hen
A female turkey.

tom
A male turkey.

Index

To Learn More

Learning more is as easy as 1, 2, 3.

1) Go to www.factsurfer.com

2) Enter "turkeys" into the search box.

3) Click the "Surf" button to see a list of websites.

With factsurfer.com, finding more information is just a click away.